Abel Hernández-Muñoz

PLANT-ANIMAL RELATIONS

Abel Hernández-Muñoz

PLANT-ANIMAL RELATIONS

Plant interactions with birds and bats

ScienciaScripts

Cover image: www.ingimage.com

This book is a translation from the original published under ISBN 978-613-9-40539-8.

Publisher:
Sciencia Scripts
is a trademark of
Dodo Books Indian Ocean Ltd. and OmniScriptum S.R.L publishing group

120 High Road, East Finchley, London, N2 9ED, United Kingdom
Str. Armeneasca 28/1, office 1, Chisinau MD-2012, Republic of Moldova, Europe
Printed at: see last page
ISBN: 978-620-7-79054-8

RE LAC ION EN

PLANT-ANIMAL

Abel Hernandez Munoz

INDEX

INTRODUCTION

Although some human groups have maintained their development without negatively altering nature, phenomena such as habitat fragmentation, the use of pesticides and the introduction of exotic species are threats mostly derived from human activities. This has led to an ecological crisis, which unfortunately has become one of the new challenges for conservation biology. It has been proposed that habitat fragmentation can modify microclimatic conditions, water and nutrient flows and light incidence, which in turn can lead to increased temperature and desiccation (Saunders et al. 1991).

It has also been suggested that fragmentation directly affects the distribution (through movements of populations and individuals) and abundance of birds (through local extinction). For example, Kattan et al. (1994) documented the local extinction of several bird species in an Andean forest several decades after fragmentation had occurred. In the same study, the authors evaluated several bird-associated characteristics, such as feeding type, vertical distribution, body and population size, degree of specialisation, and reproductive and survival rates, as possible indicators of susceptibility to extinction via fragmentation. They found that the bird species most susceptible to extinction at their study site were those at the limit of their altitudinal distribution, insectivorous understory birds and large canopy fruit-eating birds. The effects are not always detrimental, as some birds have been found to respond positively to fragmentation (Kattan et al. 1994). This is particularly expected in the case of nectarivorous, frugivorous and granivorous species that move in search of resources, especially those that move altitudinally (Loiselle & Blake 1992, Kattan et al. 1994, Ornelas

& Arizmendi 1995, Gordon & Ornelas 2000). In a pioneering study conducted in Manaus on how fragmentation affects nectarivore populations, Stouffer & Bierregaard (1995) documented that hummingbirds can respond positively in terms of abundance to the However, we did not investigate whether responses were due to variations in food resource rates between fragmentation levels. Several cases of extinction of pollinators, especially island-dwelling pollinators, have been witnessed in recent decades (Cox & Elmqvist 2000), but few contributions in the literature on how the processes are affected (Aizen & Feinsinger 1994a, Aizen & Feinsinger 1994b, Cox & Elmqvist 2000, Paton 2000). Local extinction of one of the components in plant-bird interactions can lead to extinction of the partners in the interaction. Theoretically, plant-bird interactions can be modified by processes associated with fragmentation (Rathche & Jules 1993). For example, parasitism, nest parasitism and consumption have been observed to increase as a result of fragmentation, but in none of the cases has this relationship been experimentally demonstrated. Aizen & Feinsinger (1994a, 1994b) observed that pollination levels and fruit and seed production decrease as fragment size decreases, but the magnitude of the effect is variable between fragments of the same size. This suggests that scale plays an important role in the magnitude of the effect of fragmentation on interactions. In this regard, Gordon & Ornelas (2000) have proposed that there are species that are critically dependent on a habitat, even though they apparently behave as generalists in the use of resources in that habitat ("cryptic habitat specialists"). This distinction is important as prioritisation schemes in global biodiversity conservation are established using habitat constraints (e.g. endemic species, species that are distributed in highly fragmented or diminished environments) as an indicator of ecological susceptibility. However, conservation needs may be underestimated in those regions and ecosystems that contain a high proportion of cryptic habitat specialists (e.g. altitudinal migrants). In this book we present the problems of birds that establish relationships with plants. Our aim is to describe the complexity of plant-bird interactions and their conservation. Rather than presenting an exhaustive review of the literature, we have chosen to outline the most salient points of the subject.

POLLINATION, DISPERSAL AND CONSUMPTION OF SEEDS

Although cases of obligate mutualisms are known, most plant-bird interactions are of a generalist type, both from the point of view of the plant and from the point of view of the pollinator or seed disperser. The root of this lies in opportunism on both sides. To understand this, consider what might be called the fundamental evolutionary nature of systems such as pollination and seed dispersal (Kearns et al. 1998). Plants and their pollinators and/or seed dispersers are mutualistic, each benefiting from the presence of the other. However, the resulting mutualism is not necessarily always symmetrical or cooperative. In fact, these mutualisms are thought to derive evolutionarily from relationships that were once completely antagonistic (Proctor et al. 1996). The reasons for which plants and animals establish these alliances are different, in the case of plants it is reproduction and for animals it is to obtain food. This fundamental difference can lead the interactants to a conflict of interest rather than cooperation. An obvious example is those hummingbird species that steal nectar from flowers (Ornelas 1994, Navarro 1999, Lara & Ornelas 2001). This conflict of interest dictates how natural selection will act in divergent ways for plants, on the one hand, and for those animals that plants are selectively rewarding, on the other (Kearns et al. 1998). Pollinators and seed dispersers also become agents of selection and the sexual expression of individual plants in a population will depend on their efficiency. Therefore, the evolutionary outcome of the interaction will depend on the efficiency they express in exploiting what each partner considers to be a valuable or critical resource. However, the evolutionary outcome of the interaction depends not only on the efficiency of the components, but also on the opportunism and flexibility observed in such interactions. This is important especially because it is known that the distribution of resources, in this case nectar and fruits, is not only a result of the interaction but also of other biotic and abiotic factors that determine the spatial distribution of plants. Pollinators and seed dispersers must therefore seek out and follow changes in the abundance and temporal and spatial distribution of resources (Ornelas & Arizmendi 1995). This situation makes nectarivorous and frugivorous birds more sensitive to changes in the landscape than, for example, a warbler (Parulidae). It has been postulated that pollination and seed dispersal systems may be altered after fragmentation (Paton 2000), but little information exists for the Neotropics.

Although much more information exists for pollinators and seed dispersers, flower and seed consumers also seem to follow the abundance and spatial distribution of resources (Ornelas & Arizmendi 1995, Gordon & Ornelas 2000, Renton 2001). Florivory and granivory, relatively poorly known among Neotropical birds (Poulin et al. 1994), also demand great mobility and temporal flexibility in diet on the part of birds. Several groups of birds including cracids, psittacines, columbids and emberizids consume significant quantities of flowers and seeds, but the effect they have on plant fitness has not been investigated, let alone the effects on legitimate visitors to the interaction or on the interaction as a whole. In any case, these consumers have to move at various spatial scales to exploit ephemeral and variable resources. Renton (2001) has documented such characteristics for Amazona finschi (Psittacidae). These predispersed seed-eaters are perhaps the most important seed consumers of canopy trees in the low deciduous forest of western Mexico; they can drastically decrease seed production by (1) the pattern of tree aggregation, (2) foraging typically in large groups, and (3) the long-distance movements they make daily in search of resources (Renton 2001). Thus, seed-eaters may play a fundamental role in the dynamics and maintenance of diversity in neotropical forests.

ECOLOGICAL INTERACTIONS BETWEEN PLANTS AND BIRDS

Biotic interactions involve interconnections at the community level (Jordano 1987). The local extinction of a species has consequences that are not easily predictable because in most cases detailed knowledge of the natural history is lacking. Each species on a biodiversity list adds on average two interactions in pollination systems and three in seed dispersal systems, except for alien species (Jordano 1987). These figures are underestimated because they include only mutualistic species, ignoring the contribution of antagonists (e.g. nectar stealers, florivores, seed consumers). Therefore, it is important to take care of the "permanence" of interactions over time, a s the extinction of species leads to "domino effects," which accelerate the loss of additional species. These negative effects must add up and are greater than those that would be expected as a consequence of fragmentation alone. The view that losing a plant species means losing an animal species via linked extinction, and vice versa, must be abandoned as simplistic. An integrative approach recognises that the ecological nature of plant-bird interactions lies in the richness of interconnections that vary in time and space, depending on the environmental context. To illustrate this, we describe below the interconnections that occur in a plant-bird interaction that is the subject of our research interest, suggesting that conservation biologists turn their attention to the subtleties and complexity that these systems offer.

INTERCONNECTIONS IN A PLANT-POLLINATOR SYSTEM

Heterostyly is a floral polymorphism characterised by the reciprocal position of stigmas and anthers in flowers of different plants. There are two types of heterostyly depending on whether there are two (distilia) or three (tristilia) floral morphs in the population (Barrett et al. 2000). In dystyllic plants, one of the morphs has the stigma above the anther position ("pin") and the opposite morph has the anthers above the stigma position ("thrum"). Plants with this floral dimorphism are typically pollinated by animals (Darwin 1877, Barrett 1992). Although floral morphs are supposedly designed for reciprocal transfer of pollen grains (interbreeding between morphs with reciprocal floral architecture), the efficiency of such a mechanism often depends on the effectiveness of the pollinators (Beach & Bawa 1980). If pollinators transfer pollen grains asymmetrically, reproductive success between floral morphs may be differential (Contreras & Ornelas 1999) and this may promote gender specialisation (male or female function) of floral morphs (Beach & Bawa 1980, Ornelas et al. submitted).

A population of Palicourea padifolia (Rubiaceae) near the city of Xalapa, Veracruz, Mexico, has been studied since 1996. This plant is abundant in disturbed areas of mountain mesophyll forest and is distributed from southern Mexico to Panama. Its inflorescences offer yellow tubular flowers that last only one day. This plant typically flowers from March to August (Contreras & Ornelas 1999). The population is morphologically distyllic but presents a series of asymmetries between morphs not necessarily expected in this type of polymorphism. In fact, the height of stigmas and anthers is not reciprocal and the size of corollas and stigma surface are larger in thrum than in pins (size polymorphism). Other stigmatic polymorphisms observed include the shape and size of stigmatic papillae (Ornelas et al. submitted). In addition, the anthers of pin flowers develop more pollen grains, which are smaller than those of thrum flowers (pollen heteromorphism) (Contreras & Ornelas 1999). In dystyllic plants, these asymmetries seem to play a functional role as selfing mechanisms. incompatibility and determine among other factors the efficiency of pollinators in transferring pollen grains (Hermann et al. 1999). Eleven species of hummingbirds, solitary bees and butterflies visit the flowers of P. padifolia in our study area. Non-territorial, short-billed hummingbirds are thought to be the most effective pollinators (Contreras & Ornelas 1999). Flower buds and flowers are voraciously consumed by Chlorospingus ophthalmicus (Thraupidae) and Saltator atriceps (Cardinalidae). Flower production, average nectar produced and percentage of flowers damaged by nectar robbers are equal among morphs (Contreras & Ornelas 1999), but floral parasitism by flower-ovipositing dipterans is highest among the larger flowers of the thrums. Fruits are juicy drupes, larger among pins, and produce two seeds, which are larger among thrums (Contreras & Ornelas 1999, Ornelas et al. submitted). Fruits ripen from June to December. In other study sites, fruits are consumed and possibly dispersed by Chlorospingus ophthalmicus, but little is known about frugivory and seed dispersal in this plant (Ornelas et al. submitted) (Fig. 1).

To date, floral traits commonly associated with heterostylous plants have been documented, and pollination experiments have demonstrated incompatibility between morphs (Contreras & Ornelas 1999). It is possible that these asymmetries in floral traits differentially modify the decisions of mutualistic and antagonistic visitors, depending on the resources they are seeking.

To investigate this possibility, the performance of each floral morph has been monitored for three years throughout various stages of its reproductive cycle (8 months), from inflorescence/infructescence development and production to seed survival. The working hypothesis is that the resources offered by P. padifolia (leaves, flowers and fruits) to mutualistic visitors (pollinators and seed dispersers) and antagonists (herbivores) should be equal in order to achieve symmetrical reproductive success of the floral morphs. The costs associated with such attraction were assessed in terms of herbivore attack and reproductive success, measured as size and number of fruits produced (female function), as a function of herbivory. Size asymmetries between morphs, in fruits and seeds, were evaluated in terms of germination and seedling survival.

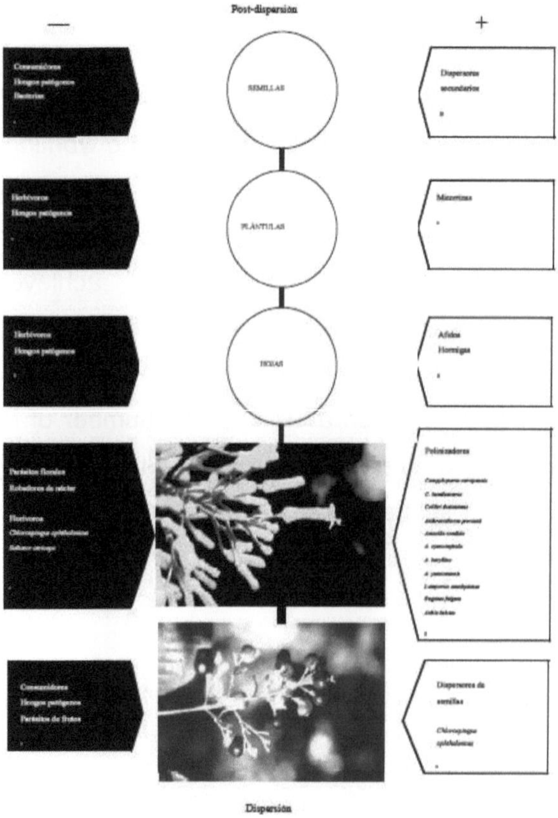

Figure 1. Diagram illustrating the complexity of the interconnections between Palicourea padifolia (Rubiaceae) and birds over the course of a year and at only one point in its geographical distribution. This plant establishes mutualistic (+) relationships with eleven species of hummingbirds (Trochilidae) and at least one frugivorous species - which presumably disperses its seeds. The results of each of these interactions are not yet known, either in time or space. However, imagine the number of possible plant-hummingbird, hummingbird-plant, hummingbird-hummingbird interactions and interconnections that may occur over the four months that plants offer nectar in their flowers to these visitors. The plant also establishes antagonistic (-) relationships with two species of birds that prey on its flowers. Although these birds consume a large number of flowers per visit, their impact on the plant's fitness is not known. The mode of consumption of these flowers by the two bird species is different, one of them ingesting the flower completely (Saltator atriceps) while the other plucks it, "sucks" it and then throws it away (Chlorospingus ophthalmicus). This has allowed us to assume that these birds belong to different compartments (florivores and nectarivores, respectively).

We found that flowering phenological performance (number of open flowers) and nectar presentation as a time reward favours pins. In contrast, fruiting phenological performance (number of fruits developed) is twice as high in thrums (Ornelas et al. submitted). Both morphs are attacked by herbivores; however, in a cafeteria experiment herbivores consumed more of the pins, which have more alkaloids. Although we do not know what this relationship means, we think that herbivores are stimulated by the presence of alkaloids. It has also been documented that the most common hummingbirds forage according to the presentation of rewards over time; however, nectar presentation is asymmetrical during the day causing pollen flow to be greater from pins to thrums. Finally, germination time and seedling survival favoured thrums under conditions of increased competition. The observed asymmetries between floral morphs of P. padifolia represent deviations from perfect heterostyly and suggest the existence of selective pressures for the evolution of genus specialisation (Ornelas et al. submitted). Our knowledge of the system is restricted to a single population. The composition and function of pollinator and seed disperser assemblages and the identity of antagonists and the magnitude of their effects may vary in time and space. We know that in this population the patterns described are maintained over the years that the population has been studied. However, it is possible that the evolutionary state of heterostyly may be different between populations of this plant, as the composition of nectarivore and frugivore assemblages is different in another population studied in Costa Rica (Ree 1997). Our working system at least illustrates the complexity of interconnections that can occur in plant-bird interactions (Fig. 1); however, we are a long way from understanding how natural and anthropogenic disturbances modify the ecological and evolutionary dynamics described here. The complexity of interconnections in the P. padifolia-hummingbird-frugivore mutualism can be addressed as a compartmentalisation problem. This term has been proposed to address energy flow in food chains (Fonseca & Ganade 1996), but can be adapted to mutualistic pollination systems (Jordano 1987, Herrera 1990, Corbet 2000). Accordingly, the interactions that occur in some systems could be seen as compartments according to the role played by each of the eleven species of hummingbirds that visit this plant. These could be formed according to the beak morphology of the hummingbirds, where species with short beaks (e.g. Atthis heloisa) and species with long beaks (e.g. Campylopterus curvipennis), remove and deposit different amounts of pollen or transfer it asymmetrically from one of the floral morphs to the other because of their differences in foraging tactics. These compartments could be further subdivided in terms of energetics; that is, hummingbirds collect different volumes of the reward according to their needs and constraints, since as the size of the hummingbird increases, so do its energetic demands. Although the degree of dependence of hummingbirds on particular plant species is generally low, some plant species may be critical to their survival (keystone species). It is likely that the presence of P. padifolia explains some of the local diversity of nectarivores and frugivores. In coffee plantations where the understorey is entirely replaced by coffee trees, the interactions described here disappear. Although it is possible that new interactions may appear in such coffee plantations, this observation may challenge the reductionist discourse of some conservationists who point out that bird diversity is maintained or is higher in some cases in coffee plantations that are managed in a friendly manner (where native trees are maintained or planted as shade for coffee). We propose that more attention needs to be devoted to those plant species on which many mutualistic (nectarivorous and frugivorous) and antagonistic (nectar-stealing) species depend, in particular the maintenance and conservation of certain vulnerable compartments.

EFFECT OF ANTAGONISTS ON INTERACTIONS

The plant-pollinator mutualism is "exploited" by species that take the rewards offered to pollinators without providing any benefit to the members of the mutualism (Bronstein 2001). Floral nectar is the most common floral reward offered to pollinators by bird-pollinated angiosperms; however, this resource is attractive to other floral visitors that steal the reward without transferring pollen between flowers (Maloof & Inouye 2000, Lara & Ornelas 2001b). Robbers typically include floral mites, bees, bumblebees, ants, birds of the order Passeriformes and even hummingbirds and bats (e.g. Ornelas 1994, Navarro 1999, Lara & Ornelas 2001b).It is known for some cases that nectar robbers have a negative effect on plant fitness and may modify foraging patterns o f pollinators. However, in some cases the effect has been documented to be positive or neutral (Maloof & Inouye 2000). Although it is difficult to generalise because these behaviours can occur facultatively (Lara & Ornelas 2001b), it has been observed that plants pollinated by hummingbirds are more sensitive to attack by nectar robbers in open areas than in the forest interior (Traveset et al. 1998). Also, plants more exposed to nectar robbers are less visited by legitimate pollinators and produce fewer seeds than those growing in the forest interior (Traveset et al. 1998). Because hummingbirds and other nectarivorous birds respond to changes in the temporal and spatial distribution of floral resources, changes in the distribution of these resources, whether natural or anthropogenic, can alter the outcomes of mutualistic and/or antagonistic plant-bird interactions in a community.

EFFECT OF DISTURBANCES ON INTERACTIONS

So far we have discussed how disturbances can affect plants and how those changes might modify the behaviours of their visitors and the outcomes of the interaction. However, birds can in turn affect the spatial distribution patterns of plants, particularly those plants that rely on vectors to disperse their propagules to the right sites and are for some reason more sensitive to changes in microenvironmental conditions. Mistletoes (Loranthaceae) are parasites that require precision seed dispersal by the vector, as seeds need to be deposited on the correct branch and host. Mistletoes are also known to be sensitive indicators of environmental conditions (water and minerals) due to their dependence on the host and birds as vectors. These characteristics make the host-vector-parasite system an excellent model to address questions of spatial distribution, where disturbance factors could be assessed at the metapopulation level and with an experimental approach.

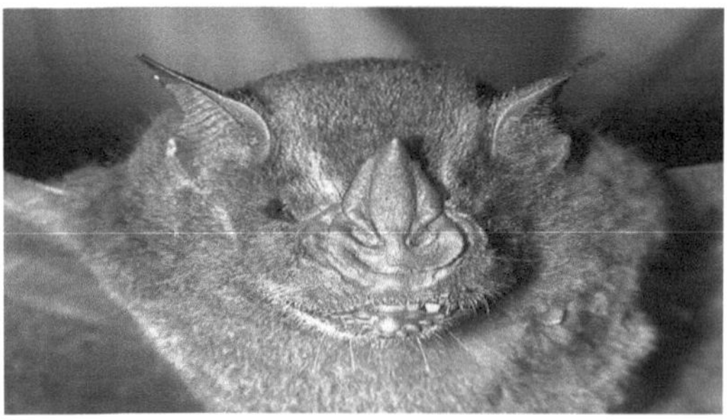

THE GEOGRAPHICAL MOSAIC AT INTERACTIONS

Thompson (1994) proposed that mutualistic interactions such as those between plants and birds are geographically structured. Thompson's proposal is based on empirical documentation suggesting that interaction outcomes for a population are spatially and temporally conditioned (Bronstein 1994a, 1994b, 2001). The outcome of a mutualistic interaction along a geographic gradient is a consequence of differences in environmental conditions along the gradient, the genetic and demographic structure of each of the interacting populations, and the biological community context in which the interaction occurs. As a result of this variation between populations, it is possible that some populations may co-evolve with their interactants in one direction and under certain selection pressures, that others may co-evolve in the opposite direction, or that in some localities one of the interactants simply does not exist. Likewise, the degree of specialisation of one population with its interactants may be very intense and close, while in other populations the interaction may be more diffuse and with a low probability of specialisation. This inter-population variation in the conditions and outcomes of interaction and the degree of specialisation that may occur creates a geographical mosaic of possibilities in plant-bird interactions. Depending on the degree of specialisation and the evolutionary trajectories and dynamics of each of the populations that make up a geographical mosaic, sites with a high potential to generate diversity ("hot spots") in all senses temporarily emerge (Fig. 2). It is obvious that maintaining such a dynamic structure requires not only a great deal of knowledge of the interaction, from its natural history to its evolutionary potential. It represents a great challenge for conservation and population management at different scales.

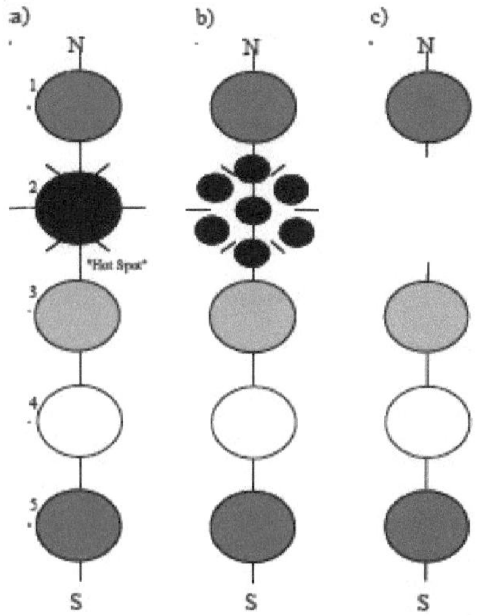

Figure 2. Possible scenarios as consequences of fragmentation in a geographical mosaic of plant-bird interactions. Imagine that a plant-bird interaction is geographically structured from north to south(a). The degree of dependence between plant and birds is influenced by local adaptations of the interactants, the degree of specialisation gradient from lower to higher (from white to black) varies between regions (1 to 5). However, there is gene flow between them (line connecting all regions). The degree of dependency and specialisation is highest in region 2, a site known in the literature as a "hotspot" because it contains a high evolutionary potential and is a generator of diversity. If this geographical region is fragmented (b), the elements in the geographical mosaic can take several directions, from the interruption of gene flow between the components. This graphical model demonstrates the importance of considering geographical scale in the conservation of plant-bird interactions.

CONSERVATION OF PLANT INTERACTIONS- AVE

Interest in the relationship between the ecology of plant-bird interactions and conservation has been minimal in any case (Janzen 1974, Kevan 1975, Howe 1984, McClanahan 1986, Feinsinger 1987). Fragmentation of natural habitats and the spread of modified environments must profoundly affect almost all aspects of plant-bird interactions (Feinsinger 1987). It has been suggested that fragmentation may affect (1) the structure and dynamics of interacting populations (Aizen & Feinsinger 1994a, 1994b) and (2) eventually the microevolution of such populations (e.g., their genetic structure). If this is true, fragmentation could be acting at the community level as a selective force for plants, as recruitment of new individuals may change by affecting the dynamics of their reproductive systems and the relationship with their pollinators and seed dispersers.

This is particularly important when thinking at the metapopulation level, where the dynamics between populations depend on how the interactants move in space. It has been suggested that pollinators are sensitive to these changes in the landscape (Bronstein 1995) and this consequently creates changes in the movement of pollen grains through space, which impact on the levels of interbreeding and reproductive success of plants. It has been postulated that the extinction of a mutualist is almost always due to drastic changes in its environment and a reduction in the number of individuals of the other component of the mutualism (Norton 1991). However, there is no information on how the process occurs and how birds react before extinction to the changes that plants undergo as a result of fragmentation. For example, it would be interesting to investigate whether the quality of the rewards that plants offer to their pollinators changes in response to changes in the landscape and how these changes alter the birds' foraging patterns and suitability.

LANDSCAPE MANAGEMENT FRAGMENTED LANDSCAPES

In heavily fragmented landscapes (e.g., only isolated trees remain), fruit-eating birds tolerant to strong fragmentation pressures have been shown to play a key role in connecting to remnant forests via seed movement (Guevara et al. 1986, McClanahan 1993, Nepstad et al. 1996). This promotes and maintains plant diversity within and between fragmented areas and facilitates the recovery of vegetation structure and species composition (Guevara et al. 1998, Medellín & Gaona 1999, Galindo-González et al. 2000). Frugivorous birds can have foraging radii of up to 25 km, providing access to gametes between plants in a foraging territory of 200-2000 km^2 (Roubik 2000). Therefore, they can play a major role in the recovery of severely fragmented landscapes by promoting the connection of landscape elements, forest fragments, regeneration cores, riparian vegetation, isolated trees and grazing areas (Galindo-Gonzalez et al. 2000). In this way, landscape components can remain connected by plants in certain pollination compartments (Corbet 2000), but eventually the system may collapse if gene flow is interrupted.

Undoubtedly, the success of any landscape management and restoration project at different scales must consider biotic interactions. Currently, reforestation programmes for landscape restoration are seen as a panacea for the environmental damage we have caused; however, there are no positive experiences to our knowledge of explicit consideration of interactions in such programmes. There is a need for more sophisticated and creative reforestation schemes to restore plant-pollinator and/or plant-seed disperser interactions (Paton 2000) and even the possibility of promoting new interactions.

HUMMINGBIRD- PLANT INTERACTIONS

Hummingbird-plant relationships have traditionally been viewed as a model of mutualism, where flowering plants, usually tubular and red in colour, produce a lot of nectar that allows them to manipulate energetically demanding visitors such as hummingbirds (Trochilidae), important in maintaining pollen flow, promoting interbreeding and genetic variability among the plant populations they visit (Feinsinger 1987, Wilding et al. 1989). The appearance of attractive traits in hummingbird plants makes them susceptible to antagonistic visitors such as bees, ants, passerine birds and even hummingbirds (Arizmendi et al. 1996, Colwell et al. 1974, Inouye 1983, Irwin & Brody 1999, Lara & Ornelas 2001, Naskrecki & Colwell 1998, Ornelas 1994, Paciorek et al. 1995), which collect rewards without offering pollination services. The arrival of unwanted nectar-stealing visitors is not the only risk that arises in this mutualism. Hummingbirds can act as vectors of floral mites (Colwell 1973, Lara & Ornelas 2001) and pathogenic fungi (Lara & Ornelas 2001) that directly and/or indirectly affect both hummingbirds and their host plants. Transmission of these organisms can lead to sexually transmitted diseases (Jennersten 1983, Roy 1993, Roy 1994), pollen consumption (Paciorek et al. 1995), decreased nectar volume (Colwell 1995, Lara & Ornelas 2001) and changes in the sex ratio between flowers of protandric plants (Lara & Ornelas 2001). Therefore, the harmony of the hummingbird-plant mutualist may be modified by the effect of these antagonists.The presence of mites on plants pollinated by hummingbirds has been well documented (Colwell 1979, Dobkin 1987, Heyneman et al. 1991, Naeem et al. 1985). But the effect among interactants has been little explored (Colwell 1995, Lara & Ornelas 2001). Flower mites (Acari: Mesostigmata: Ascidae) feed on pollen and nectar of a wide variety of plant species pollinated exclusively by hummingbirds (Colwell 1995, Lara & Ornelas 2001, Paciorek et al. 1995). To disperse to new flowers, the mites climb into the beak of a visiting hummingbird and travel on the nostrils until a new visit is made to another inflorescence of the appropriate host species (Colwell 1985, Dobkin 1990) (Fig. 1). The relationship between floral mites and their host hummingbirds has typically been defined as phoretic (Colwell 1973, Dobkin 1990, Lara & Ornelas 2001); however, it has been suggested that transport itself may represent an energetic cost to hummingbirds (Colwell 1995) and it is not known whether mites produce any damage within nostrils (Lara & Ornelas 2001). Nectar consumption by mites may reduce its availability to hummingbirds (Colwell et al. 1974, Kearns et al. 1998), indirectly affecting the reproductive success of the host plant (Lara & Ornelas 2001, Paciorek et al. 1995). Nectar volume in flowers of Moussonia deppeana (Gesneriaceae), a protandric plant pollinated by the hummingbird Lampornis amethystinus, decreases by up to 50% in the presence of flower mites (Tropicoseius sp. nov.) without affecting seed production (Lara & Ornelas 2001). However, nectar consumption by mites may influence the foraging patterns of visiting hummingbirds and this may indirectly affect pollen transmission (Lara & Ornelas 2001). The negative effect of floral mites on nectar volume seems to be a widespread phenomenon among hummingbird-pollinated plants whose flowers last several days and produce large volumes of nectar (Lara & Ornelas 2001). Certain plant pathogenic fungi such as rusts, blights and endophytic fungi can use plant pollinators to disperse their spores (Batra 1987, Bultmant & White 1988, Jennersten 1983, Lara & Ornelas 2001, Pfunder & Roy 2000, Roy 1993, Wilding et al. 1989). These fungi have invaded structures and processes used by plants to attract pollinators, through strategies such as floral mimicry ("pseudoflowers") and the hijacking of plant reproductive structures. This interaction potentially affects pollination ecology, plant life history traits, floral evolution, as well as the dynamics of pathogen transmission and evolution (Roy 1994).

Infections by pathogenic fungi, such as those attacking anthers, have generally been studied in plants pollinated by insects. It is interesting to explore whether fungi that attack anthers of flowers visited by hummingbirds use them for dispersal and/or alter their host in a way that benefits the fungus. The pathogenic fungus Fusarium moniliforme Sheld (Deuteromycota) can systematically infect flowers in the male (staminate) phase of M. deppeana, replacing pollen with microconidia and ascospores, which are dispersed to other flowers by hummingbirds, thus spreading the infection (Lara & Ornelas 2001). This suggests that the effects of F. moniliforme on the interaction are negative (reduction in success reproductive stage of their host plants), as the flowers in the male phase are virtually sterilised. Knowledge generated to date shows that pollination by animals is essential for the sexual reproduction of most higher plants. Accelerated habitat alteration by human activity, together with the accelerated invasion of alien species and the expectation of global climate change, which threaten plants and pollinators both phenologically and ecologically, have led in recent years to the conception of conservation of plant-pollinator interactions (Kearns et al. 1998). The results shown here suggest that knowledge about antagonistic organisms that exploit such relationships should be considered in conservation plans.

INTERACTIONS AT THE LANDSCAPE LEVEL

The maintenance of landscape-level processes that vary in time and space are essential for the long-term conservation of biodiversity (Harris et al. 1996). However, anthropogenic disturbances, such as immoderate logging and land-use changes, currently affect interactions, many of them critical, such as those between plants and their pollinating or dispersing birds (Howe & Smallwood 1982, Kearns et al. 1998). Forest fragmentation and increased edge zones have been suggested as major factors in the increase of parasitic and epiphytic plant populations in disturbed areas (Norton et al. 1995, Williams-Linera 1992). These plants are dispersed either passively by air or directly by vectors, which are mostly birds that deposit seeds in sites suitable for germination and establishment (Davidar 1983, Sargent 1995, Wheelwright et al. 1984). When dispersal is mediated by birds, as in the case of parasitic plants, the search for host trees and the probability that healthy individuals will become infected depends more on the abundance of parasitized individuals in the population than on the absolute density of host trees (Antonovics et al. 1993, Martínez del Río et al. 1996). This so-called "frequency-dependent" mode of transmission can lead to highly unstable population dynamics that are easily modified by disturbances (Martínez del Rio et al. 1996), which can affect not only the interactions of parasitic plants with the birds that pollinate or disperse them, but also interactions with species that prey on them (Norton & Reid 1997).Many ecological and evolutionary processes in ecosystems derive from plant-parasite interactions because of the importance of hosts as primary producers (Gilbert & Hubbell 1996). In healthy ecosystems, infections by persistent parasites such as mistletoes usually affect low numbers of individuals and play a key role in community dynamics. In disturbed ecosystems, infections can become epidemic and affect large numbers of individuals, causing permanent changes to the forest community (Gilbert & Hubbell 1996).Mistletoes (Loranthaceae) are hemiparasitic flowering plants that obtain water and mineral resources from their perennial host plants (Burger & Kuijt 1983, Kuijt 1969). Their fruits constitute important nutritional resources for the birds that disperse them and numerous associations have been documented between the abundance of their fruits and the abundance of frugivorous birds (Reid et al. 1995, Restrepo 1987).In the case of the mistletoe Psittacanthus schiedeanus, it has been observed that maximum fruit production, with a high lipid content, occurs during the winter period when other resources are scarce, in order to attract the greatest possible number of dispersing birds (López de Buen & Ornelas 2001). In central Veracruz, Mexico, the main dispersing bird species (16 bird species have been observed consuming its fruits) are the grey or florican Capulineros (Ptilogonys cinereus), the Chinitos (Bombycilla cedrorum) and the gregarious Luises (Myiozetetes similis), which influence the abundance and distribution of mistletoe during their foraging activities (López de Buen & Ornelas 1999). It is thought that these birds may recognise the most infested host trees in search of fruit to feed on. In fact, when they arrive at a group of trees, they first visit the individuals with the most mistletoe plants and from there fly to nearby trees, parasitized or not, to rest or consume the fruits previously collected. Thus, these birds disperse the whole, viable seeds, separated from the rest of the fruit, during their defecation or regurgitation on the high branches of the host trees (López de Buen & Ornelas 1999).In the P. schiedeanus study, it was observed that only 35% of the fruits were removed, while the rest of them remained on the plants of origin, where they dried out or fell to the ground and were attacked by fungi or consumed by non-dispersing species (López de Buen & Ornelas 2001). Although birds help to remove seeds from mistletoe plants to their new host trees, only a small fraction of these seeds will manage to germinate and establish as sexually adult plants (López de Buen & Ornelas 2001) and subsequently produce fruits with "infecting" seeds.Fragmentation processes in forested areas are often so rapid that the plant and animal communities that inhabit them often do not have sufficient time to adapt to these disruptive changes (Rodhes & Odum 1996). In mistletoes, like in many other plants heliophytes and opportunists in exploiting disturbed environments with high light intensity, the increase in the extension of border areas can promote the increase of populations and infestation rates (López de Buen & Ornelas 2001, López de Buen et al. 2001, Martínez del Río et al. 1996), as long as the degree of fragmentation and disturbance of the environment is not too severe. On the contrary, in cases of extreme disturbance, population declines may be promoted by the disappearance of possible interactions with birds that pollinate them or disperse their fruits (Norton & Reid 1997).

CLIMATE CHANGE AND PLANT-BIRD INTERACTIONS

In the last decade, climate change (CC) has increasingly occupied the scientific literature. Here we are concerned with how much and how CC may affect plant-bird interactions and how this possible effect may influence bird conservation.The main changes in this process are the 0.3-0.6 °C increase in mean annual temperature over the last 100 years and the increase in atmospheric carbon concentration, creating different global warming scenarios (IPCC 1996, McCarty 2001). Changes in recent decades are apparent at all levels of ecological organisation: changes in population, life history and distribution of organisms, changes in species composition and changes in ecosystem structure and functioning (McCarty 2001). The possible future implications for species and community conservation have been addressed in several papers (IPCC 1996, Sala et al. 2000, McCarty 2001).For plants, the main effects of CC include, among many others, increases in primary productivity, changes in photosynthetic rate (differentials for C3, C4 and CAM species), biases in species distribution limits, changes in germination and recruitment, and changes in community structure and dynamics (Melillo et al. 1990, Prentice 1992, IPCC 1996, Ceulemans et al. 1999, Bradley et al. 1999, Ni et al. 2000, Loisseau & Soussana 2000, NRC 1999, McCarty 2001). For birds, the main changes detected are biases in geographic distribution and changes in breeding date and nesting habitat selection (Crick & Lennon 1999, Thomas & Lennon 1999, Shaeter et al. 2000, Martin 2001, Moss et al. 2001, McCarty 2001). Very little is known about the effect of CC on biotic interactions (Kareiva et al. 1993, IPCC 1996, NRC 1999, Hughes 2000), with aspects such as plant-herbivore interactions having been addressed (Coviella & Trumble 1999). The importance of plant-bird interactions in ecosystem functioning has been amply demonstrated and a hypothetical example of the effects of CC on their dynamics and their relationship with bird conservation in the Americas can be considered. For example, a reserve in an area of cloud forest, intended to preserve endemic birds, is located between 700 and 1700 masl. Studies at the reserve show that the frequency and abundance of species from lower altitudes is increasing, and that they begin nesting two weeks earlier than they did in the last decade. The researchers report that the reproductive success of endemic birds has declined significantly due to a decrease in the availability of insects, which are important for chick development. In addition, the most important insects for nestling diet are mostly consumed by birds typical of lower altitudes, insects that are also floral visitors to plants whose cover has decreased in the last twenty years. Further studies report that the lower altitudinal limit of endemic birds has risen by 300 m and that the plants on which the insects that are the staple diet of their nestlings depend do not follow this pattern of altitudinal change, which is associated with changes in soil moisture and temperature that determine the recruitment of these plants.In addition, among endemic birds, two species of hummingbirds have not been recorded for three years and the nectar production of the plants mostly visited by hummingbirds is negatively affected by the decrease in relative air humidity, the annual average of which has decreased by 5% in the last twelve years. Hummingbirds have one of the highest metabolic rates among birds and therefore high energy requirements, which they obtain in part from floral nectar. The researchers conclude that unless climate trends are reversed, the endemic bird species most sensitive to changes in food availability will become locally extinct. That is, the conservation importance of the reserve has been diminished by the magnitude of CC.

The effects of anthropogenic disturbances can lead to changes such as habitat fragmentation, which can interact with CC, producing synergistic effects on organisms. For example, at the landscape level, in forested systems the loss of tree cover leads to an increase in albedo (refraction of the sun's rays by the earth's surface) (IPCC 1996), enhancing the effects of climate change originating at a larger spatial scale. The CC and other problems we face in the conservation of plant-bird interactions draw attention to the need to broaden the temporal and spatial scales of study. On the other hand, the perspective of the geographical mosaic of co-evolution (Thompson 1994, 1997) suggests that the dynamics of interactions must be analysed at the geographical scale, as they vary in time and space, generating patterns that are difficult - if not impossible - to discern at the local scale.We cannot yet answer how CC affects, with what magnitude and in what direction, plant-bird interactions, but we can see how small changes in climate can increase the variance of response of species and their interactions, so we should be open to finding innovative ways to investigate and examine the conservation problems of plant-bird interactions.

PIONEER PLANTS IN THE DIET OF BIRDS AND BATS IN THE GUAMUHAYA MOUNTAINS, CUBA

INTRODUCTION

One of the essential characteristics of tropical ecosystems is the relationship between frugivores and plants. Among the fauna inhabiting the tropics, birds, bats and flightless mammals are considered to be the most responsible for the dispersal of a large number of plant species (Fleming et al., 1987). The high mobility of these vertebrates maintains and increases plant genetic variation, as it facilitates the interbreeding of unrelated individuals, and increases the likelihood that offspring of different parents grow together and exchange genes during pollination (Begon et al., 1986). The forests of the Biosphere Reserve "Sierra del Rosario" present a high structural complexity and floristic diversity (Menéndez et al., 1988). These forests produce a large number of fruits adapted to be consumed and disseminated by vertebrates. Given the absence of the main families of neotropical frugivorous birds (Cotingidae, Pipridae, Ramphastidae and Steotornithidae) as well as primates, among other frugivorous terrestrial mammals (e.g. Procionidae); phyllostomid bats and non-specialised frugivorous birds represent the main consumers of fruits and therefore constitute essential elements for the natural restoration of the forested areas of the Cordillera. Birds are a conspicuous group within the Reserve's forests, both in terms of diversity and biomass, and 92 species have been reported to date (Rodríguez et al., 1999). Although there are no specialised frugivores, it is known that many Cuban bird species frequently include fruits in their diets (Kirkconnell et al., 1992); however, little is known about the functional role of this important group of vertebrates in the forested regions of Cuba. On the other hand, bats of the family Phyllostomidae, endemic to the American tropics, are responsible for seed dispersal of hundreds of plant species, including epiphytes, trees and shrubs (Gardner, 1977). Numerous studies (e.g. Foster et al., 1986; Fleming, 1988) have shown that bats play an important role in colonisation by pioneer plants of disturbed habitats, which modify abiotic conditions and biotic environment, and allow the establishment of other primary plant species. All phytophagous phyllostomid species have been captured in the Sierra del Rosario Reserve. As part of the project "Regenerative strategies and application of pre-germinative treatments on seeds of pioneer forest species in the Guamuhaya Mountains, Cuba" we set out to find out which pioneer plants form part of the diet of birds and bats that inhabit the Cordillera.

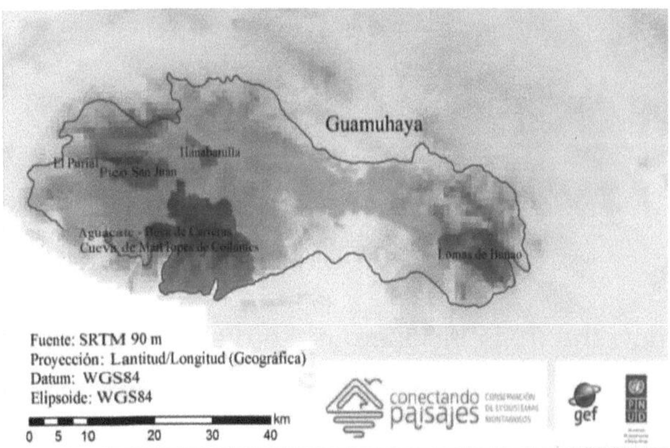

MATERIALS AND METHODS

To find out which bird species make use of pioneer plants, observations were made on at least five fruiting adult individuals of the following plant species: Muntingia calabura, Cecropia schreberiana, Piper aduncum, Trema micrantha, Trichospermun mexicanum, Guazuma ulmifolia and Talipariti elatum (formerly Hibiscus elatus). Observations were made monthly, in the morning hours, as this is the period of greatest bird feeding activity, on selected trees in the vicinity of the Ecological Station "Sierra del Rosario" (N 220 51' 04.2" W 820 55' 52.8"). The species of birds, which regularly include fruits in their diets, were recorded and special attention was paid to those observed consuming them. In addition, all anecdotal data related to fruit consumption by birds in other localities of the Reserve were compiled.

In order to know the fruits included in the diet of the phyllostomid bats, captures were made in forest patches of El Aguacate, Manantiales, Pico San Juan, El Naranjo, Lomas de Banao and El Garrote. In the first two localities two samplings were carried out (rainy and dry periods), in the last one only the rainy season was monitored. The bats were captured with mist nets (9 and 12 m x 2.5 m) placed at ground level. A total of five to six nets were used, kept open from 18:00 to 24:00 hrs, during five consecutive nights. The animals captured were identified and marked with numbered metal rings. They were then placed in cloth bags for one hour to give them time to defecate and collect faeces. The food samples were labelled with the number of each animal and placed in plastic vials with 70% alcohol for later identification in the laboratory. In the case of pollinivorous bats, pollen samples were taken by passing small cubes of gel over the body (Beattie, 1971). Another method used to identify the plants used as food by the bats was the collection of fruit remains found under the diurnal and feeding roosts.

RESULTS AND DISCUSSION

A total of 19 bird species, included in nine families (Table 1), were the most common in the selected trees of the pioneer plant species. The most representative families were: Columbidae with four species, and Icteridae and Picidae with three. With the exception of the Cat Thrush, Dumetella carolinensis, all breed in Cuba and are known to include plant elements in their diets (Kirkconnell et al., 1992). Three species of woodpeckers (Family Picidae) were highly frequent, being observed on six of the seven plant species monitored. Other birds with an equal number of plant taxa visited were the Red-crowned Woodpecker, Priotelus temnurus; the Song Thrush, Turdus plumbeus; and the Yellow-billed Woodpecker, Agelaius humeralis. Of the plants, Trema micrantha and Muntingia calabura with 17 and 16 bird species respectively showed the highest diversity of visitors (Table 1). Both species have been previously reported in the diet of birds (Snow, 1981). Piper aduncum, with five bird species, showed the lowest number of visitors. This plant has a bushy habit (probably too fragile for medium-sized birds) and the position of its fruits does not seem to favour its consumption by birds, although the genus has been reported in the diet of non-specialised frugivorous birds in the Neotropics (Snow, 1981). The pioneer plants included in the observations have generally small fruits, of the drupes and berries type, with small seeds; these characteristics have been pointed out as adaptations of these plants to be consumed by non-specialist birds (Snow, 1971). On two occasions, individuals of the Arriero, Saurothera merlini, were detected feeding on the fruits of Guazuma ulmifolia, although not in the specimens selected for this study. Despite the lack of fleshy fruits, the consumption of Trichospermun mexicanum and Talipariti elatum by some bird species was observed (Table 1). A Negrito, Melopyrrha nigra, was also observed consuming fruits of Hibiscus costatus; prior to this work, the genus Hibiscus had been reported as part of the diet of birds in Africa (Snow, 1981). In addition to the pioneer plant species, the main objective of this study, other species that birds frequently include in their diet were observed, such as Copey (Clusia rosea; Clusiaceae), Macurije (Matayba apetala; Sapindaceae) and Almácigo (Bursera simaruba, Burseraceae). The fruits of the Copey were consumed by Green Woodpecker, Xiphidiopicus percussus; Jabado Woodpecker (Melanerpes superciliaris), Bien-te-Veo (Vireo altiloquus) and Solibio (Icterus dominicensis); those of the Macurije were consumed by Bien-te-Veo, Jabado Woodpecker, Green Woodpecker, Tocororo and Totí (Dives atroviolacea); while the Almácigo was predated by Tocororo, Solibio, Bien-te-Veo and Black-throated Blue Jay (Dendroica caerulescens). A group of nine individuals of the San Diego Bluebell's Warbler (Cyanerpes cyaneus) was observed feeding on the Copey. Other plants included in the diet of the Cordillera birds were Palo de Caja, Allophylus comina (Sapindaceae) and Aguedita, Picramnia pentandra (Simaroubaceae), both consumed by the Tocororo; the Cigua, Nectandra coriacea (Lauraceae), by the King Thrush; the Yaba, Andira inermis (Fabaceae) by the Solibio and the Macagua, Pseudolmedia spuria (Moraceae) by the Chillina, Teretristis fernandinae. Table 1. Most common birds that visited pioneer plants in the Biosphere Reserve "Sierra del Rosario". Mc: Muntingia calabura; Cs: Cecropia schreberiana; Pa: Piper aduncun; Tm: Trema micrantha; TRm: Trichospermum mexicanum; Gu: Guazuma ulmifolia; Te: Talipariti elatum. P: perching on the plant; N: feeding on the flowers; C: consuming the fruit.

	Ma	Cs	Pa	Tm	TRm	Gu	Te
COLUMBIDAE							
Columba leucocephala	P	-	-	P	P	-	-
Columba squamosa	P	-	-	P	P	-	-
Zenaida asiatica	-	P	-	-	-	P	P
Zenaida macroura	-	P	-	P	P	P	-
TROGONIDAE							
Priotelus temnurus	P	P	P	P, C	P, C	-	P
PICIDAE							
Colaptes auratus	P	P	-	P	P	P	P
Melanerpes supercilii	P	P	-	P	P	P	P, C
Xiphidiopicus percussus	P	P	-	P	P	P	P
TYRANNIDAE							
Tyrannus caudifasciatus	P	-	P	P	-	P	P
Tyrannus dominicensis	P	-	P	P	-	P	P
TURDIDAE							
Turdus plumbeus	P	P	-	P	P	P	P
MIMIDAE							
Dumetella carolinensis	P	-	-	P	-	-	-
Mimus polyglottos	P	-	-	P	-	P	-
ICTERIDAE							
Agelaius humeralis	P	P	-	P	P	P	P, C
Dives atroviolacea	-	P	-	-	P	P	P
Icterus dominicensis	P, C	-	-	P, C	P	-	P. N
THRAUPIDAE							
Cyanerpes cyaneus	P, N	P	-	P, N	P	P	P, N
Spindalis zena	P, C	P	P	P, C	P	P	P
EMBERIZIDAE							
Melopyrrha nigra	P, C	P	P	P, C	-	P	-

All the bird species observed can be classified as facultative frugivores as they supplement their diets with elements of animal origin. Despite being a conspicuous group in the plants studied, pigeons should not be considered as frugivorous birds given the characteristics of their feeding habits, essentially granivorous, which causes the destruction of seeds, so they are not involved in their dispersal. A total of 391 individuals of phyllostomid bats were captured, of which 90% corresponded to three species: Monophyllus redmani, Artibeus jamaicensis and Phyllonycteris poeyi. Samples of feeding activity were obtained from 35.5% of the individuals (Table 2). Seeds of three pioneer plant species (Cecropia schreberiana, Piper aduncum, and Muntingia calabura) were the most frequent items in the samples. Pollen from another pioneer plant, Talipariti elatum, was also found to be an essential element for nectar-eating bats in the dry season. When available, the fruits of Syzygium jambos, Ficus sp., Guazuma ulmifolia, Solanum umbellatum, Sideroxylon foetidissimum, Andira inermis and Callophyllum antillanum. Little is known about the feeding habits of Phyllops falcatus; in addition to the species listed in Table 2, it is known to include C. schreberiana and S. jambos in its diet (Mancina and García, 2000).

Phyllonycteris poeyi, Brachyphylla nana and Erophylla sezekorni, despite being considered nectarivorous bats, are known to include fruits in their diets (Silva Taboada, 1979). Analysis of the faeces of P. poeyi corroborates the importance of fruits as part of their diet, as 35% of the species' troops had seeds in their faeces, all of them from understorey plants. It seems that this resource is more important during the rainy season, as 46% of the individuals captured during this period contained seeds, mostly of Piper aduncum and Muntingia calabura (Table 2). This may correspond to the lower availability of flowers during the rainy season. During the dry season, the most common element in the P. poeyi samples was pollen, mainly from Hibiscus sp (Malvaceae). Monophyllus redmani was found to include small fruits in its diet, three samples contained small seeds of Trema micrantha. Before this work it was not known that M. redmani in Cuba included fruits in its diet. As in P. poeyi, pollen is the most frequent element in the samples from the dry period.Based on the floristic list of the Guamuhaya Mountains (Herrera et al., 1988) and a review of the available literature, it was found that in the Cordillera there are at least 148 infrageneric taxa, included in 62 plant families, which are reported as part of the diet of birds and phyllostomid bats in the Neotropical region. The most represented families were: Rubiaceae (10), Myrtaceae (6), Mimosaceae(6), Lauraceae (6) and Euphorbiaceae (6), Sapindaceae (6), and Moraceae. Birds seem to preferentially include species of the families Rubiaceae, Lauraceae, and Euphorbiaceae; and bats Mimosaceae, Myrtaceae and Caesalpiniaceae. Fleming (1979) noted that both groups exploit different types of fruit and that competition is low. It is likely that the species of Ficus (Moraceae) present in the Cordillera, and reported in the literature as part of the diet of birds, are consumed by bats. As part of this work, Artibeus jamaicensis have been captured on several occasions carrying fruits of Ficus sp. and remains have been found under their diurnal roosts, but unfortunately these have not been identified. Artibeus jamaicensis, together with Phyllops falcatus, are the Cuban species most specialized in the consumption of fruits, which they collect directly from the foliage of trees and shrubs.

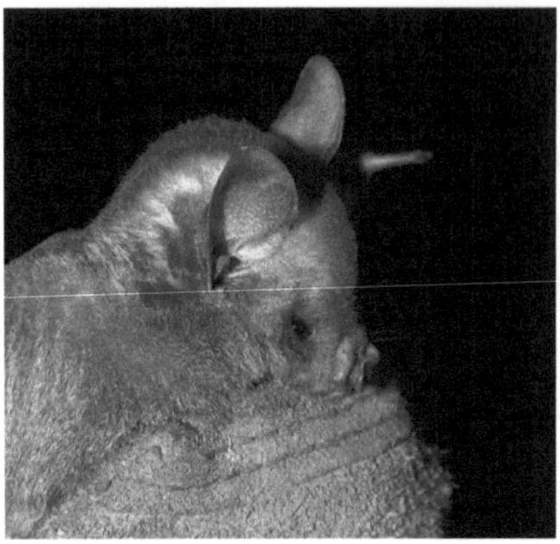

Table 2. Plant species found as part of the diet of phyllostomid bats in the Guamuhaya Mountains according to the time of collection (dry and rainy). N: number of individuals collected.

SPECIES	N	SECA	RAIN
Artibeus jamaicensis	116	Piper aduncum (2)	Cecropia schreberiana(16) Piper aduncum (2) Unidentified seeds(2)
Phyllops falcatus	15	Piper aduncum (1) Muntingia calabura (1)	Piper sp. (1)
Monophyllus redmani	122	Trema micrantha (1) Unidentified seeds (1) POLLEN Spathodea campanulata (3) Talipariti elatum (20) Hibiscus rosasinencis (1) Hibiscus sp. (6) Ochroma pyramidale (1)	Trema micracantha (2)
Erophylla sezekorni	1	Unidentified pollen (1)	
Phyllonycteris poeyi	116	Piper aduncum (2) Conostegia xalapensis (2) Unidentified seeds (2) Lantana trifolia (1) POLLEN Spathodea campanulata (9) Talipariti elatum (19) Hibiscus rosasinencis (1) Hibiscus sp. (9)	Piper aduncum (18) Remains of mesocarp (2) Mutingia calabura (7) Remains unidentified (3)
Brachyphylla nana	21	Ficus sp. (1) Muntingia calabura (1)	Unidentified seeds (1)

From analysis of the faeces of A. jamaicensis, Cecropia schreberiana was found to be an important element in its diet, at least in the rainy period. Other plants detected in the faeces of A. jamaicensis were Piper aduncum and from observations of food remains in diurnal roosts and foraging activity it was determined that this species consumes the diet of many species of neotropical bats (Kalko, et al., 1996) and in Cuba. Silva Taboada (1979) reports four species in the diet of A. jamaicensis. From research in continental forest areas of the Neotropics, it is known that there are differences in the rainfall of small seeds between vertebrate groups of birds and bats. The latter contribute more to seedfall in open areas, whereas bats contribute more to seedfall in open areas, whereas bats contribute more to seedfall in open areas. birds deposit seeds around fruiting trees, undergrowth and edges of forest patches (Foresta, et al., 1984; Gorchov et al., 1993). In the West Indies, studies related to the functional role of vertebrates as seed dispersers are very scarce. Future research is needed to understand the differential role of birds and bats in the natural regeneration of affected areas of the Guamuhaya Mountains.

ACKNOWLEDGEMENTS

To the workers of the Guamuhaya Mountains for their hospitality during this work. To all colleagues who supported us during the night work sections. To Lic. Alicia Rodríguez, researcher at the National Botanical Garden for the identification of the pollens. To Corinna Koch and WildLife Preservation Trust Canada for their support of our research.

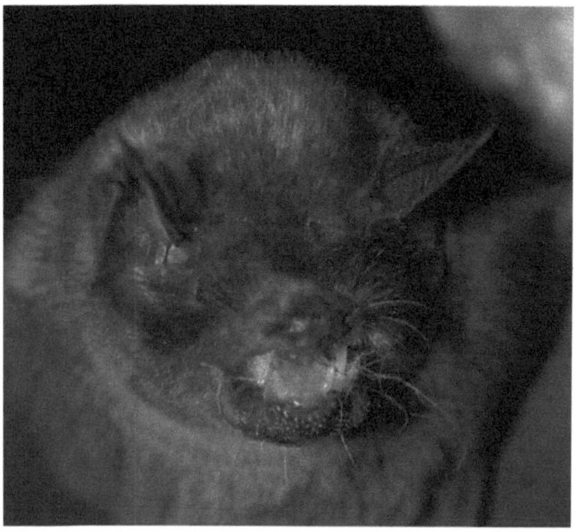

REFERENCES ON FRUGIVORÍA

Beattie, A. J. 1971. A technique for the study of insect-borne pollen. Pan-Pacific Entomol. 47:82.

Begon, M.E., L. Harper and C.R. Townsend. 1986. Ecology Individuals, Populations, and Communities. Blackwell Scientific Publications, Oxford. 269 pp.

Fleming, T. H. 1979. Do tropical frugivores compete for food ?. Amer. Zool. 19: 1157-1172.

Fleming, T. H. 1988. The Short-tailed Fruit Bat. A study in plant-animal interactions. The University of Chicago Press. 365 pp.

Fleming, T. H., R. Breitwisch, and G. H. Whitesides. 1987. Patterns of tropical vertebrate frugivore diversity. Ann. Rev. Ecol. Syst. 18: 91-109.

Foresta, H. de, P. Charles-Dominique, C. Erard, and M. Prevost. 1984. Zoocorie et premiers stades de la régénération naturelle après coupe en forêt guyanaise. Rev. Ecol., 39: 369- 400.

Foster, R. B., J. Arce and T. S. Wachter. 1986. Dispersal and the sequential plant communities in Amazonian Perufloodplain. In: A. Estrada and T. H. Fleming (eds.)Frugivores and seed dispersal. Dordrecht: Junk. pp 151-172.

Gardner, A. L. 1977. Feeding habits. In: R. J. Baker, J. K. Jones and D. C. Carter (eds.) Biology of bats of the New World family Phyllostomatidae. Carter (eds.) Biology of bats of the New World family Phyllostomatidae. Part II. Special Public. The Museum Texas Tech Lubbock, No. 13. pp. 293-350.

Gorchov, D. L., F. Cornejo, C. Ascorra, and M. Jaramillo. 1993. The role of seed dispersal in the natural regeneration of rain forest after strip-

cutting in the Peruvian Amazon. Vegetatio 107/108: 339-349.
Kalko, E. K., E. A. Herre, and C. O. Handley Jr. 1996. O. Handley Jr. 1996. Relation of fig fruit characteristics to fruit-eating bats in the New and Old world tropics. J. Biogeography, 23: 565-576.
Kirkconnell, A; O. H. Garrido, R. M. Posada, and S. O. Cubillas. 1992. Trophic groups in avifauna.Cuban. Poeyana, 415, 1-21.
Menéndez, L., E. E. García, R. A. Herrera, M. E. Rodríguez and J. A. Bastart. 1988. Structure and productivity of the middle evergreen forest of the Sierra del Rosario, Chap 8, 151-211 pp. In: R. A. Herrera, L. Menendez, M.E. Rodríguez and E. E. García (eds.). Ecology of Evergreen Forests of the Sierra del Rosario, Cuba. MAB Project no.1, 1974-1987. ROSTLAC, Montevideo Uruguay.
Mancina, C. A., A. Hernández Marrero, L. Rodríguez Schettino. 2000. Wild mastofauna of the "Sierra del Rosario" Biosphere Reserve, Cuba. Poeyana 476-480:9-13.
Mancina, C. A. and L. García. 2000. Notes on the natural history of Phyllops falcatus (Gray, 1839) (Phyllostomidae: Stenodermatinae) in Cuba. Chiroptera Neotropical 6 (1-2): 123-125.
Rodríguez, L. S., C. A. Mancina, E. Pérez, A. Hernández, and A. Chamizo. 1999.
Management and conservation of
terrestrial vertebrates of the Guamuhaya Mountains, as a basis for studying climate change. Final Project Report. CITMA, Havana.
Silva Taboada, G. 1979. The Bats of Cuba. Editorial Academia. Havana, Cuba. 423 pp.
Snow, D. W. 1971. Evolutionary aspects of fruit-eating by birds. Ibis 113: 194- 202.Snow, D. W. 1981. Tropical frugivorous birds and their food plants: A world survey. Biotropica 13(1): 1-14.

CONCLUSIONS

The slow rate at which information is being generated on the natural history of plant-animal interactions, in particular, scientifically verifiable data on pollinator and/or seed disperser abundance and rigorously identified effects of processes such as fragmentation on interactions, represents a serious threat to conserving pollination and/or seed dispersal systems in the face of time pressures, resource availability, and increasing rates of deforestation and fragmentation. Pollination and seed dispersal systems must be studied on their own terms, including differences among individuals in foraging experiences (Bronstein 1995, Ornelas & Arizmendi 1995), abundance or scarcity of nest sites, extent of foraging areas (Renton 2001), flowering peaks and opportunistic strategies (Lara & Ornelas 2001b). Therefore, it is necessary to pay attention to how changes in the landscape promote changes in plant reproductive success, flower and fruit selection, quality of rewards and services, seed germination, seedling survival and migratory routes (Roubik 2000).In conclusion, we believe that the conservation or restoration of plant-bird interactions requires an understanding of how each system reacts to multiple interconnected factors. In addition to Kremen & Ricketts (2000), we believe that future approaches to the study of plant-bird interactions and their conservation should consider (1) the phenology of interactions between plants and their pollinators and/or seed dispersers across space, (2) the role of antagonists and three-level interactions, (3) the relationship between fragmentation and availability of nest sites and food resources (flowers and fruits),(4) the level of redundancy within compartments, (5) the vagility of pollinators, seed dispersers and consumers and their ability to move through unfavourable environments, (6) the identification of key resources in native environments and those generated in anthropogenically produced landscapes, and (7) the inclusion of sensitivity measures in management and conservation programmes across time and space and at different scales. This approach requires a monumental effort across scales, landscapes and boundaries. Only then can solutions be implemented This will allow us to assess the role of flower and seed consumers and antagonists such as nectar stealers and/or pathogens in a holistic way in their context (Buchmann & Nabhan 1996) and analogously to that of seed dispersers. In the Antilles, studies related to the functional role of vertebrates as seed dispersers are very scarce. Future research is needed to understand the differential role of birds and bats in the natural regeneration of affected areas in the mountains of Cuba.

BIBLIOGRAPHY

Aizen, M. A. & Feinsinger, P. 1994a. Forest fragmentation, pollination and plant reproduction in a Chaco dry forest, Argentina. Ecology 75: 330-351.

Aizen, M. A. & Feinsinger, P. 1994b. Habitat fragmentation, native insect pollinators, and feral honeybees in Argentine "Chaco Serrano". Ecol. Appl. 4: 378-392.

Antonovics, J., P. H. Thrall, A. M. Jaroz & D. Stratton. 1993. Ecological genetics of metapopulations: the Silene-Ustilago plant pathogen system. Pp. 146170 in
L. Real (ed.). Ecological genetics. Princeton University Press, Princeton. New Jersey, USA.

Arizmendi, M. C., C. A. Domínguez & R. Dirzo. 1996. The role of an avian nectar robber and of hummingbird pollinators on the reproduction of two plant species. Funct. Ecol. 10: 119-127.

Barrett, S. C. H. 1988. Heterostylous genetic polymorphisms: model system for evolutionary analysis. Pp. 1-24 In S. C. H. Barrett (ed.). Evolution and function of heterostyly. Springer-Wien. New York.

Barrett, S. C. H., D. H. Wilken & W. W. Cole. 2000. Heterostyly in the Lamiaceae: the case of Salvia brandegeei. Plant Syst. Evol. 223: 211-219.

Batra, S. W. T. 1987. Deceit and corruption in the blueberry patch. Nat. Hist. 96: 57-59.

Beach, J. H. & K. S. Bawa.1980. Role of pollination in the evolution of dioecy from distantly. Evolution 34: 1138-1142.

Bradley, N. L., A. C. Leopold, J. Ross & W. Huffaker. 1999. Phenological changes reflect climate change in Wisconsin. Proc. Nat. Acad. Sci. USA 96: 9701-9704.

Bronstein, J. L. 1994a. Conditional outcomes of mutualistic interactions. Trends Ecol. Evol. 9: 214-217.

Bronstein, J. L. 1994b. Our current understanding of mutualism. Quarterly Rev. Biol. 69: 31-51.

Bronstein, J. L. 1995. The plant/pollinator landscape. Pp. 256-288 In L. Fahrig,

L. Hansson & G. Merriam (eds.). Mosaic landscapes and ecological processes. Chapman and Hall. New York.

Bronstein, J. L. 2001. The exploitation of mutualisms. Ecology Letters 4: 277287.

Buchmann, S.. L. & G. P. Nabhan. 1996. The forgotten pollinators. Island Press. Washington, D.C.

Bultmant, T. L. & J. F. White. 1988. Pollination of a fungus by a fly. Oecologia 73: 317-319.

Burger, W. & J. Kuijt. 1983. Loranthaceae sensu lato. Flora Costaricensis. Feldiana Bot. 13: 29-79.

Ceulemans, R., L. A. Janssens & M. E. Jach. 1999. Effects of CO2 enrichment on trees and forests: lessons to be learned in view of future ecosystem studies. Ann. Bot. 84: 577-590.

Colwell, R. K. 1973. Competition and coexistence in a simple tropical community. Am. Nat. 107: 737-760.

Colwell, R. K. 1979. The geographical ecology of hummingbird flower mites in relation to their host plants and carriers. Pp. 461-468 In J. G. Rodriguez (ed.). Recent advances in Acarology, Vol. 2. Academic Press. New York.

Colwell, R. K. 1985. Stowaways on the hummingbird express. Nat. Hist. 94: 56-63.

Colwell, R. K. 1995. Effects of nectar consumption by the hummingbird flower mite Proctolaelaps kirmsei on nectar availability in Hamelia patens. Biotropica 27: 206-217.

Colwell, R. K., B. J. Betts, P. Bunnell, F. L. Carpenter & P. Feinsinger. 1974. Competition for the nectar of Centropogon valerii by the hummingbird Colibri thalassinus and the flower-piercer Diglossa plumbea, and its evolutionary implications. Condor 76: 447-452.

Contreras, P. S. & J. F. Ornelas. 1999. Reproductive conflicts of Palicourea padifolia (Rubiaceae) a distylous shrub of a tropical cloud forest in Mexico. Plant Syst. Evol. 219: 225-241.

Corbet, S. A. 2000. Conserving compartments in pollination webs. Conserv. Biol. 14: 1229-1231.

Coviella, C. E. & J. T. Trumble. 1999. Effects of elevated atmospheric carbon dioxide on insect-plant interactions. Conserv. Biol. 13: 700-712.

Cox, P. A. & T. Elmqvist. 2000. Pollinator extinction in the Pacific Islands. Conserv. Biol. 14: 1237-1239.

Crick, H. Q. P. & T. H. Sparks. 1999. Climate change related to egg-laying trends. Nature 399: 423-424.

Darwin, C. 1877. The different forms of flowers on plants of the same species. John Murray. London, United Kingdom.

Davidar, P. 1983. Birds and neotropical mistletoes: effects on seedling recruitment. Oecologia 60: 271-273.

Dobkin, D. S. 1987. Synchronous flower abscission in plants pollinated by hermit hummingbirds and the evolution of one-day flowers. Biotropica 19: 90- 93.

Dobkin, D. S. 1990. Distribution patterns of hummingbird flower mites (Gamasidae: Ascidae) in relation to floral availability on Heliconia inflorescences. Behav. Ecol. 1: 131-139.

Feinsinger, P. 1987. Approaches to nectarivore-plant interactions in the New World. Revista Chilena Hist. Nat. 60: 285-319.
Feinsinger, P. 1987. Approaches to nectarivore-plant interactions in the New World. Revista Chilena Hist. Nat. 60: 285-319.
Galindo-González, J., S. Guevara & V. J. Sosa. 2000. Bat- and bird-generated seed rains at isolated trees in pastures in a tropical rainforest. Conserv. Biol. 14: 1693-1703.

Gilbert, G. S. & S. P. Hubbell. 1996. Plant diseases and the conservation of tropical forest. BioScience 46: 98-106.

Gordon, C. E. & J. F. Ornelas. 2000. Comparing endemism and habitat restriction in the Mesoamerican tropical deciduous forest avifauna: implications for biodiversity conservation planning. Bird Conserv. Int. 10: 289-304.

Guevara, S., S. Purata & E. van der Maarel. 1986. The role of remnant trees in tropical secondary succession. Vegetatio 66: 74-64.

Harris, L. D., T. S. Hoctor & S. E. Gergel. 1996. Landscape processes and their significance to biodiversity conservation. Pp. 319- 347 In O. R. Rodhes, R. K.

Chesser & M. H. Smith (eds.). Population dynamics in ecological space and time. University of Chicago Press. Chicago, Illinois, USA.

Hermann, B. P., T. K. Mal, R. J. Williams & N. R. Dollahon. 1999. Quantitative evaluation of stigma polymorphism in a tristylous weed, Lythrum salicaria (Lythraceae). Am. J. Bot. 86: 1120-1129.

Herrera, C. M. 1999. Daily patterns of pollinator activity, differential pollinating effectiveness, and floral resource availability, in a summer-flowering Mediterranean shrub. Oikos 58: 277-278.

Heyneman, A. J., R. K. Colwell, S. Naeem, D. S. Dobkin & B. Hallet. 1991. Host plant discrimination: experiments with hummingbird flower mites. Pp. 455- 485 In P. W. Price, T. M. Lewinsohn, G. W. Fernandes & W. W. Benson (eds.). Plant-animal interactions: evolutionary ecology in tropical and temperate regions. Wiley-Interscience. New York.

Howe, H. F. 1974. Implications of seed dispersal by animals for tropical reserve management. Biol. Conserv. 30: 261-281.

Howe, H. F. & J. Smallwood. 1982. Ecology of seed dispersal. Annu. Rev. Ecol. Syst. 13: 201-228.

Hughes, L. 2000. Biological consequences of global warming: is the signal already apparent? Trends Ecol. Evol. 15: 56-61.

Inouye, D. W. 1983. The ecology of nectar robbing. Pp. 153-173 In B. Bentley & T. Elias, (eds.). The biology of nectaries. Columbia University Press. New York.

IPCC (Intergovernmental Panel on Climate Change). 1996. Climate change 1995: the science of climate change. Contribution of working group I to the second assessment report of the IPCC. Cambridge University Press. Cambridge, Massachusetts.

Irwin, R. E. & A. K. Brody. 1999. Nectar-robbing bumblebees reduce the fitness of Ipomopsis aggregata (Polemoniaceae). Ecology 80: 1703-1712.

Janzen, D. H. 1974. The deflowering of Central America. Nat. Hist. 83: 48-53.

Jennersten, O. 1983. Butterfly visitors as vectors of Ustilago violacea spores between caryophylaceous plants. Oikos 40: 125-130.

Jordano, P. 1987. Patterns of mutualistic interactions in pollination and seed dispersal: connectance, dependence asymmetries, and coevolution. Am. Nat. 129: 657-677.

Kareiva, P. M., J. G. Kingsolver & R. B. Huey (eds.). 1993. Biotic interactions and global change. Sinauer Associates. Sunderland, Massachusetts.

Kattan, G. H., H. Alvarez-López & M. Giraldo. 1994. Forest fragmentation and bird extinctions: San Antonio eighty years later. Conserv. Biol. 8: 138-146.

Kearns, C. A., D. W. Inouye & N. M. Waser. 1998. Endangered mutualisms: the conservation of plant-pollinator interactions. Annu. Rev. Ecol. Syst. 29: 83- 112.

Kevan, P. G. 1975. Pollination and environmental conservation. Environ. Conserv. 2: 293-298.

Kremen, C. & T. Ricketts. 2000. Global perspectives on pollination disruptions. Conserv. Biol. 14: 1226-1228.

Kuijt, J. 1969. The biology of parasitic flowering plants. University of California Press. Los Angeles. California.

Lara, C. & J. F. Ornelas. 2001. Preferential nectar robbing of flowers with long corollas: experimental studies of two hummingbird species visiting three plant species. Oecologia 128: 263-273

Lara, C. & J. F. Ornelas. submitted. Amethyst-throated Hummingbird Lampornis amethystinus responses to experimental simulation of hummingbird flower mite nectar consumption on Moussonia deppeana (Gesneriaceae). Oikos.

Lara, C. & J. F. Ornelas. 2001. Nectar "theft" by hummingbird flower mites and
its consequences for seed set in Moussonia deppeana. Funct. Ecol. 15: 78-84.

Lara, C. & J. F. Ornelas. submitted. Nectar consumption by flower mites on six hummingbird-pollinated plants with contrasting flower longevities. Ecology.

Lara, C. & J. F. Ornelas. in prep. Hummingbirds as vectors of Fusarium moniliforme spores to the protandric shrub Moussonnia deppeana: taking advantage of a mutualism.

Loiselle, B. A. & J. G. Blake. 1992. Population variation in a tropical bird community: implications for conservation. BioScience 42: 838-845.

Loisseau, P. & J. F. Soussana. 2000. Effect of elevated CO2 , temperature and N fertilization on nitrogen fluxes in a temperate grassland ecosystem. Global Ecol. Biogeogr. Letters 6: 953-965.

López de Buen, L. & J. F. Ornelas. 1999. Frugivorous birds, host selection and the mistletoe Psittacanthus schiedeanus, in Central Veracruz, Mexico. J. Trop. Ecol. 15: 329-340.

López de Buen, L. & J. F. Ornelas. submitted. Host compatibility of the cloud forest mistletoe Psittacanthus schiedeanus (Loranthaceae) in

Central Veracruz, Mexico. Am. J. Bot.

López de Buen, L. & J. F. Ornelas. 2001. Seed dispersal of the mistletoe Psittacanthus schiedeanus by birds in Central Veracruz, Mexico. Biotropica.

López de Buen, L., J. F. Ornelas & J. G. García-Franco. 2001. Mistletoe infection of trees located at fragmented forest edges in the cloud forest of Central Veracruz, Mexico. J. Forest Ecol. Manag.

Maloof, J. E. & D. W. Inouye. 2000. Are nectar robbers cheaters or mutualists? Ecology 81: 2651-2661.

Martin, T. E. 2001. Abiotic vs. biotic influences on habitat selection of coexisting species: climate change impacts? Ecology 82: 175-188.

Martínez del Rio, C., A. Silva, R. Medel & M. Hourdequin. 1996. Seed dispersers as disease vectors: bird transmission of mistletoe seeds to plant hosts. Ecology 77: 912-921.

McCarty, J. P. 2001. Ecological consequences of recent climate change. Conserv. Biol. 15: 320-331.

McClanahan, T. R. 1986. Pollen dispersal and intensity as criteria for the minimum viable population and species reserves. Environ. Manag. 10: 381- 386.

McClanahan, T. R. 1993. Accelerating forest succession in a fragmented landscape and the role of birds and perches. Conserv. Biol. 7: 279-288.

Medellín, R. & O. Gaona. 1999. Seed-dispersal by bats and birds in forest and disturbed habitats in Chiapas, Mexico. Biotropica 31: 478-485.

Melillo, J. M., T. V. Callaghan, F. I. Woodward, E. Salati & S. K. Sinha. 1990. Effects on ecosystems. Pp. 283-310 In Houghton, J.T., G. J. Jenkins & J. J. J.

Ephraums (eds.). Climate change. The IPCC scientific assessment. Cambridge University Press. Cambridge, Massachusetts.

Moss, R., J. Oswald & D. Baines. 2001. Climate change and breeding success: decline of the capercaillie in Scotland. J. Anim. Ecol. 70: 47-61.

Naeem, S., D. S. Dobkin & B. M. Oconnor. 1985. Lasioseius mites (Acari: Gamasina: Ascidae) associated with hummingbird-pollinated flowers in Trinidad, West Indies. Int. J. Ento. 27: 338-353.

Naskrecki, P & R. K. Colwell. 1998. Systematics and host plant affiliations of hummingbird flower mites of the genera Tropicoseius Baker and Yunker and Rhionoseius Baker and Yunker (Acari: Mesostigmata: Ascidae). Thomas Say Publications in Entomology. Monographs, Entomological Society of America.

National Research Council. 1999. Global Environmental Change: Research Pathways for the Next Decade. Committee on Global Change Research.

Navarro, L. 1999. Pollination ecology and effect of nectar removal in Macleania bullata (Ericaceae). Biotropica 31: 618-625.

Nepstad, D. C., C. Uhl, C. A. Pereira & J. M. Cardoso da Silva. 1996. A comparative study of tree establishment in abandoned pasture and mature forest of eastern Amazonia. Oikos 76: 25-39.

Ni, J., M. T. Sykes, I. C. Prentice & W. Cramer. 2000. Modelling the vegetation of China using the process-based equilibrium terrestrial biosphere model BIOME3. Global Ecol. Biogeogr. Letters 9: 463-479.

Norton, D. A. 1991. Trilepidea adamsii: an obituary for a species. Conser. Biol. 5: 52-57.

Norton, D. A. & N. Reid. 1997. Lessons in ecosystem management from management of threatened and pest Loranthaceous mistletoes in New Zealand and Australia. Conserv. Biol. 11: 759-769.

Norton, D. A., R. J. Hobbs & L. Atkins. 1995. Fragmentation, disturbance and plant distribution: mistletoes in woodland remnants in the Western Australia wheatbelt. Conserv. Biol. 9: 426-438.

Ornelas, J. F. & M. C. Arizmendi. 1995. Altitudinal migration: implications for conservation of avian Neotropical migrants in western Mexico. Pp. 98-112 In
M. H. Wilson & S. A. Sader (eds.). Conservation of Neotropical migratory birds in Mexico. Maine Agricultural and Forest Experiment Station. Miscellaneous Publication 727, Orono, Maine.

Ornelas, J. F. 1994. Serrate tomia: an adaptation for nectar robbing in hummingbirds? Auk 111: 703-710.

Ornelas, J. F., C. González, L. Jiménez, C. Lara, A. J. Martínez & P. S. Contreras (submitted). Ecological factors around heterostyly in hummingbirdpollinated Palicourea padifolia (Rubiaceae). Am. J. Bot.

Paciorek, C. B., B. Moyer, R. Levin & S. Halpern. 1995. Pollen consumption by the mite Proctolaelaps kirmsei and possible effects on Hamelia patens. Biotropica 27: 258-262.

Paton, D. C. 2000. Disruption of bird-plant pollination systems in Southern Australia. Conserv. Biol. 14: 1232-1234.

Pfunder, M. & B. A. Roy. 2000. Pollinator-mediated interactions between a pathogenic fungus, Uromyces pisi (Pucciniaceae), and its host plant, Euphorbia cyparissias (Euphorbiaceae). Am. J. Bot. 87: 48-55.

Poulin, B., G. Lefebvre & R. McNeil. 1994. Characteristics of feeding guilds and variation in diets of bird species of three adjacent tropical sites. Biotropica 26: 187-197.

Prentice, I. C. 1992. Climate change and long-term vegetation dynamics. Pp. 293-345 In Glen-Lewin, D. C., R. K. Peet & T. T. Veblen (eds.). Plant succession. Theory and prediction. Chapman & Hall. New York.

Proctor, M., P. Yeo & A. Lack. 1996. The natural history of pollination. Timber. Portland, Oregon.

Rathcke, B. J. & E. S. Jules. 1993. Habitat fragmentation and plant-pollinator interactions. Current Science 65: 273-277.

Ree, R. H. 1997. Pollen flow, fecundity, and the adaptive significance of heterostyly in Palicourea padifolia (Rubiaceae). Biotropica 29: 298-308.

Reid, N., M. Stafford & Z. Yan. 1995. Ecology and population biology of mistletoes. Pp. 285- 311 In M. D. Lowman & N. M. Nadkarni (eds.). Forest canopies. Academic Press. New York.

Renton, K. 2001. Lilac-crowned parrot diet and food resource availability: resource tracking by a parrot seed predator. Condor 103: 62-69.

Restrepo, C. 1987. Ecological aspects of the dissemination of five mistletoe species by birds. Humboldtia 1: 65-116.

Rodhes, O. R. & E. P. Odum. 1996. Spatiotemporal approaches in ecology and genetics: the road less travelled. Pp. 1- 7. In O. R. Rodhes, R. K. Chesser & M. H. Smith (eds.). Population dynamics in ecological space and time. The University of Chicago Press. Chicago, Illinois.

Roubik, D. W. 2000. Pollination system stability in tropical America. Conserv. Biol. 14: 1235-1236.

Roy, B. A. 1993. Floral mimicry by a plant pathogen. Nature 362:56-58.

Roy, B. A. 1994. The use and abuse of pollinators by fungi. Trends Ecol. Evol. 9: 335-339.

Sala, O. E., F. Stuart Chapin III, J. J. Armesto, E. Berlow, J. Bloomfield, R. Dirzo,

E. Huber-Sanwald, L. F. Huenneke, R. B. Jackson, A. Kinzig, R. Leemans, D.
M. Lodge, H. A. Mooney, M. Oesterheld, N. LeRoy Poff, M. T. Sykes, B. H. Walker, M. Walker & D. Wall. 2000. Global biodiversity scenarios for the year 2100. Science 287: 1770-1774.

Sargent, S. 1995. Seed fate in a tropical mistletoe: the importance of host twig size. Funct. Ecol. 9: 197-204.

Saunders, D., R. Hobbs & C. Margules. 1991. Biological consequences of ecosystem management: a review. Conserv. Biol. 5: 18-32.

Shaeter, B. E., J. Tufto, S. Engen, K. Jerstad, O. W. Restad & J. E. Skatan. 2000. Population dynamical consequences of climate change for a small temperate songbird. Science 287: 854-856.

Stouffer, P. C. & R. O. Bierregaard. 1995. Effects of forest fragmentation on understory hummingbirds in Amazonian Brazil. Cons. Biol. 9: 1085-1094. Thomas, C. D. & J. J. Lennon. 1999. Birds extend their ranges northwards. Nature 399: 213.

Thompson, J. N. 1994. The coevolutionary process. University of Chicago Press. Chicago.

Thompson, J. N. 1994. The coevolutionary process. University of Chicago Press. Chicago, Illinois.

Thompson, J. N. 1997. Evaluating the dynamics of coevolution among geographically structured populations. Ecology 78: 1619-1623.

Traveset, A., M. F. Willson & C. Sabag. 1998. Effect of nectar-robbing birds on fruit set of Fuchsia magellanica in Tierra del Fuego: a disrupted mutualism. Funct. Ecol. 12: 459-464.

Wheelwright, N. T., W. A. Haber, K. G. Murray & C. Guindon. 1984. Tropical fruit-eating birds and their food plants: a survey of a Costa Rican lower montane forest. Biotropica 16: 173-192.

Wilding, N., N. M. Collins, P. M. Hammond & J. F. Webber. 1989. Insect-fungus interactions. Academic Press. New York.

Williams-Linera, G. 1992. Distribution of the hemiepiphyte Oreopanax capitatus at the edge and interior of a Mexican lower montane forest. Selbyana 13: 35-38.

Withgott, J. 1999. Pollination migrates to top of conservation agenda. BioScience 49: 857-862.

FSC
www.fsc.org
MIX
Papier aus verantwortungsvollen Quellen
Paper from responsible sources
FSC® C105338

Printed by Books on Demand GmbH, Norderstedt / Germany